I0803745

IMAGES
of America

BIRMINGHAM BROADCASTING

A true celebrity is one who needs to use only a single name—Cher, Liberace, Elvis, Madonna, Annette. WAPI-TV's ever-effervescent Rosemary required no further identification for viewers. (Rosemary Lucas collection.)

On the Cover: "Slick" and "sophisticated" were not two words that described local television news in the early days. Channel 13's newscast in 1953 included, from left to right, Bart Darby at the weatherboard, Charlie Davis, and Dan Daniel. (Alvin Hudson collection.)

IMAGES
of America

BIRMINGHAM BROADCASTING

Tim Hollis

ARCADIA
PUBLISHING

ISBN 978-0-7385-4271-3

Published by Arcadia Publishing
Charleston, South Carolina

Printed in the United States of America

Library of Congress Catalog Card Number: 2006922051

For all general information contact Arcadia Publishing at:
Telephone 843-853-2070
Fax 843-853-0044
E-mail sales@arcadiapublishing.com
For customer service and orders:
Toll-Free 1-888-313-2665

Visit us on the Internet at www.arcadiapublishing.com

The new WBRC-TV studios, completed atop Red Mountain in September 1954, boasted a commanding view of downtown Birmingham and giant neon letters that could be seen for miles. (Alvin Hudson collection.)

CONTENTS

INTRODUCTION

Why has there never been a book documenting the rich history of radio and television broadcasting in Birmingham? That is a very good question, and after you have sampled the multitude of different topics in this one, it may be a bit easier to understand the answer. A comprehensive history of Birmingham broadcasting would require several hundred pages, so for now, we will just hit the high spots and hope that someone is inspired to give the topic a more in-depth treatment at some point in the future.

In a few of my previous books, I briefly detailed my own interest in the subject of Birmingham television, but for those who came in late I will give a capsule version here. My television debut came about on the occasion of my fifth birthday in 1968, when my parents booked me for an appearance on WBRC's *Birthday Party* show, hosted by news anchor Joe Langston. It's too bad this occurred in March rather than during a sweeps period, because the television debut of a five-year-old author would certainly have helped WBRC make a big jump in the ratings for that particular program. As it turned out, Channel 6 was usually a ratings leader anyway because of its combination of formidable talent and powerful signal before cable television, so it obviously didn't need my help. I was undaunted; it would not be my last appearance on that station over the ensuing decades.

Most of the Birmingham kids' shows had a minimum age limit for their guests, and because I was still a tiny tot, I never got to appear on any of the others until the early 1970s. By that time, Neal Miller's *Sgt. Jack Show* was the only one left on the air, and I made the steep trip up Golden Crest Drive to the WBMG studios at least a couple of times for that. I had a great desire to become a puppeteer, and the crazy characters created by Miller's assistants, Howard Cruse and Ted Lowry, were a big influence.

While I was in college, I received a call from one of those television legends whom I had never met in person, Cousin Cliff Holman, who was looking for a puppeteer to make personal appearances with him, so I had at last come full circle to where I had been 10 years earlier. When Cliff's television series was revived as *Cousin Cliff's Clubhouse* on WBRC in 1990–1993, I was on hand as the puppeteer behind the desk. About the time that show debuted, I wrote my first book, which was a biography titled *Cousin Cliff: 40 Magical Years in Television*. That, I suppose, was the beginning of this book, although many other books of mine on varied topics have appeared between the two.

Many of you have probably seen my previous book for Arcadia, *Birmingham's Theater and Retail District*. That one touched briefly on Birmingham's radio and television history as it crossed over with the larger picture of the department stores and movie houses of downtown. Likewise, in this volume, you will find the veteran downtown businesses making a few guest appearances in the story of broadcasting in the city. There is one major difference between these two subjects, however. Because the coverage area of the Birmingham television stations (and especially the radio stations) was so large, many people who never even approached downtown Birmingham in person could still listen to or watch Joe Rumore, Pat Gray, Wendell Harris, Joe Langston, Rosemary Lucas, Cousin Cliff, and Benny Carle.

So just set your dial to this station and prepare for a trip back in time. We will begin by listening in on the radio, then we shall see how television came to town in 1949. We'll check in on the news, sports, and weather and chuckle along with our favorite kids' show hosts. After a word from some of the best-known sponsors in town, we will wrap up with the talk shows and variety shows that enlivened the Birmingham airwaves. Our program will begin in just a moment, but first we must acknowledge the real sponsors—those individuals who contributed their time, memories, photographs, and other artifacts to make this book so visually appealing. In alphabetical order, let's give a big hand to these friends: Carol Aldy, Evelyn Allen, Rodney Barstein, Keith Barze, Greg Bass, Jim and Vicki Battan, Rice Baxter, Lisa Belcher, David Black, Bill Bolen, Dick Breit, Dan Brennan, Steling Brewer, Country Boy Eddy Burns, Benny Carle, Lanny Chappelear, Kyle Clark, Barry Copeland, Don Dailey, Bart Darby, Joe Dentici, Pat Dietlin, Margo George, Frank Giardina, Greg Glenn, Pat Gray, Courtney Haden, Edie Hand, Pat Barnes Harmer, Wendell Harris, Everett Holle, Cousin Cliff Holman, Jane Hooper, Todd Jones, Ernest Langner, Pat Linton, Jim Lucas Jr., Rosemary Lucas, Don Martin, John McGiboney, Ward McIntyre, Chris Mitchell, Ed Moran, Chris Campbell Oyer, Jean Pierce, Donnie Pitchford, Patsy Preston, Gary Rigsby, Tom Roberts, Dave Roddy, James Robert Romine, Joanna Sasnett, Errick Smith, M. D. Smith III, Phil Stewart, Julie Strauss, J. D. Weeks, Russell Wells, June West Wetzel, Beth Willauer, Keith Williams, Glenn Wills, Roy Wood, Bill Wright, and Tom York.

The author made his first visit to a Birmingham television station at age five in March 1968. The cedar tree on the WBRC front lawn was the centerpiece of an annual Christmas lighting ceremony during those years. (Author's collection.)

One

Turn Your Radio On

The beginning of radio broadcasting in Birmingham reads a bit like alphabet soup, but just hang on to your antenna, and maybe we can get it straightened out. The first attempt to put a Birmingham radio station on the air occurred in April 1922, when Alabama Power Company signed on with its new WSY. The station failed to make any sort of impression, even after moving its studios into the elegant Loveman's department store, so in 1925, Alabama Power cut off the juice and donated WSY's equipment to station WMAV, which was operating out of the campus of Alabama Polytechnic Institute in Auburn. This new combination of WSY and WMAV became known as WAPI, after the university's initials, but for the next several years its reign would remain mainly down on the plains.

Meanwhile, back in Birmingham, the first successful station was WBRC (for Bell Radio Corporation), which signed on the month after WSY left town. WBRC was soon joined by WKBC in February 1926, but as we shall soon see, that station became much better known under a different name. Finally on December 31, 1928, WAPI moved from Auburn to Birmingham, giving the city three large stations and setting a pattern that would survive for decades.

While the 1930s and 1940s are rightfully known as the "golden age of radio" because of the wealth of dramatic and comedy programming on the national networks, the story of local radio hardly gets started until after World War II. The loss of network programming, due to television's arrival, caused local stations to really and truly develop their own personality announcers and disc jockeys. Beginning in the late 1940s, a steady stream of familiar names began streaming over the Birmingham airwaves: Joe Rumore and his "home folks" style that appealed to many and irritated a few; Joe's more bombastic brother, Duke, who had much the same effect; Neal Miller, the girls' heartthrob; Jim Lucas, who even brushed with national fame for a while; Tommy Charles, who began as a singer but later evolved into one of Birmingham radio's most durable comedians and satirists; those cheerful rivals Happy Hal Burns and Happy Wilson; and the list goes on and on.

Let's now tune our dial and listen a while and find out what sort of radio sounds we can pull out of the accumulated static. First let's see if we can pick up anything on what was once the most popular spot on the dial—a frequency known as "the Big 610."

Birmingham's second successful radio station, WKBC, signed on in February 1926, and its studios (above) and transmitter (below) were located in the Tutwiler Hotel. In April 1934, WKBC was purchased by the *Birmingham News/Age-Herald* and changed its name to WSGN (for South's Greatest Newspaper). WSGN would subsequently go on to become Birmingham's most legendary of the old-time stations. (Alvin Hudson collection.)

After moving its studios around town for several years, WBRC finally landed at the corner of Second Avenue North and Eighteenth Street. It remained here until the entire block, which also contained the Ritz Theater, was demolished in 1982. (Ernest Langner collection.)

WBRC's transmitter site was on U.S. Highway 78 at East Thomas, and it is still used by its successor, WERC. In 1941, Dixie Neon installed this impressive sign at the tower; notice the figure of an announcer holding a microphone at the top of the right hand post. (Ernest Langner collection.)

Playing off the popularity of Robert Ripley's *Believe It or Not!*, in 1937 WAPI featured this weekly series highlighting the bizarreness in Alabama. (Rice Baxter collection.)

From the late 1940s until 1953, WSGN operated out of this building on Red Mountain. As a forerunner to WBRC-TV's sign, the neon letters could be seen from all over downtown. (Ernest Langner collection.)

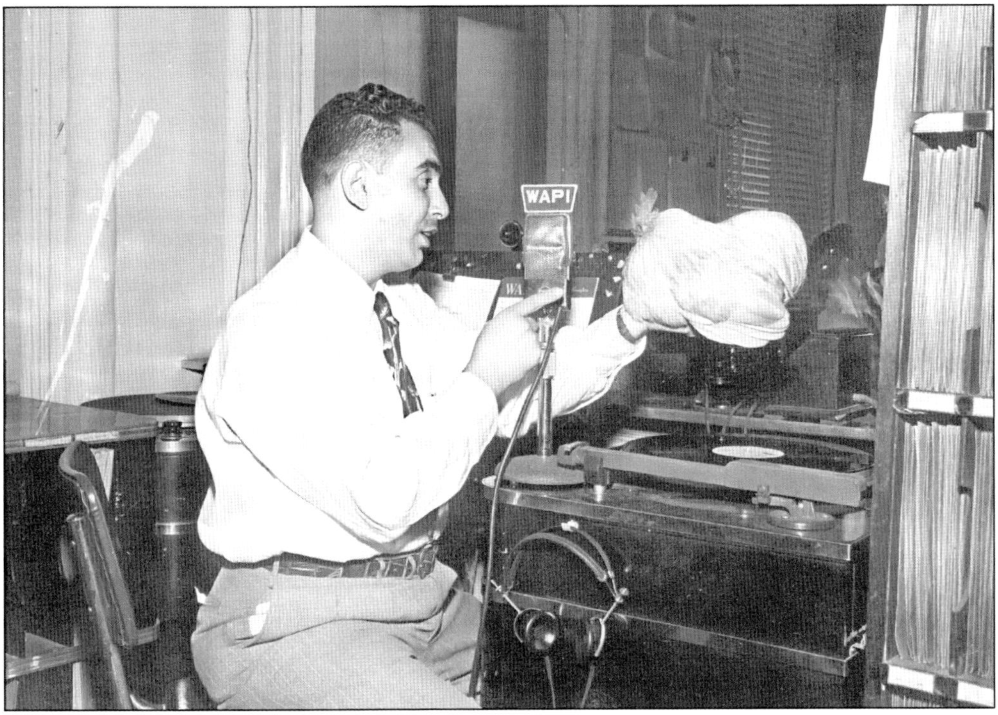

During his days with WAPI in 1947, Joe Rumore appears to be demonstrating that he isn't chicken by deliberately trying to make his show lay an egg. (Alvin Hudson collection.)

WTNB went on the air in June 1946 as a Mutual Network affiliate. Over the years it would evolve first into WILD, then into today's WYDE. (Alvin Hudson collection.)

One of the most fondly remembered WSGN features was the *Comic Club*, on which "Miss Ann" read the comic strips out of the station's parent company's newspapers. Although there were several Miss Anns over the years, this is the one who remained with the show the longest, teacher Evelyn Walker of Ramsay High School. (Author's collection.)

With *The Cat Mountain Boys*, WSGN seemed to be lifting an idea or two from the network comedy serial *Lum and Abner*. Announcer Jim Cleary looks in on the general store high jinks of Effie, Amos, Orville, Grandpappy, and other rustic denizens. (Author's collection.)

WVOK went on the air in October 1947 and soon established its studios on the Bessemer Super Highway. Its 50,000 watts made it a superpower among Birmingham's older stations. (Todd Jones collection.)

FM radio was ahead of its time in the late 1940s. WBRC-FM built this studio atop Red Mountain in 1947, but it was soon converted to television use instead. (Alvin Hudson collection.)

After Joe Rumore took his down-home folksy style to WVOK in 1948, the two were a perfect fit for each other, and he remained the station's flagship personality for the rest of his career. (Patsy Preston collection.)

WVOK was not always a country station; in 1958 they were playing rock and roll and sponsored this stage show starring teen heartthrob Fabian. Joe Rumore is apparently giving the young sprout some sage advice about radio. (Patsy Preston collection.)

WBRC's Happy Hal Burns was happier than ever in October 1950, when he posed with Yolande Betbeze of Mobile, the first (but not last) Miss Alabama to win the national title of Miss America. (Alvin Hudson collection.)

Women's features on WBRC were handled by British-accented Mary Badham. Her daughter, also named Mary Badham, later co-starred as Scout with Gregory Peck in *To Kill a Mockingbird*. (Alvin Hudson collection.)

This stunning panorama shows the crest of Red Mountain rapidly developing into "Radio Park," as it was unofficially known. The tiny building to the right of the tower is fledgling television Channel 13, which would move into the WSGN building seen here in 1953. The stores, theaters, and office buildings of downtown loom to the north. (Alvin Hudson collection.)

As if things were not hilarious enough with Happy Hal Burns on WBRC, in 1953, WAPI was offering the music of Happy Wilson and his group. Ted Crabtree is handling the keyboard while "Professor" Huel Murphy tinkles the ivories; later *Grand Ole Opry* singer Mary Ann Worth and Happy Wilson himself are at the microphone. (Alvin Hudson collection.)

Dave Campbell was one of WAPI's most famed announcers. He was reportedly the first in Birmingham to host a call-in show. *The People Speak* was the precursor to today's glut of talk radio stations. (Alvin Hudson collection.)

Maury Farrell handled a multitude of duties for WAPI and moved effortlessly into television when the station began Channel 13 in 1949. (Alvin Hudson collection.)

Neal Miller was a giant on WSGN in the 1950s—literally, as he stood 6 feet 7 inches tall. As we shall see, in the 1960s he became the beloved "Sergeant Jack" of children's television fame. (Jean Spradlin-Miller collection.)

WAPI's Sterling Brewer could be found doing everything from anchoring the news to hosting a children's program to emceeing the Saturday night wrestling matches. (Rosemary Lucas collection.)

Bill Bolen began his Birmingham career at WSGN in 1954 and more than a half-century later was still anchoring the news on Channel 6. (Bill Bolen collection.)

Among the teenage crowd, you were squaresville unless you were seen parked outside WSGN's Sky Castle broadcast booth at Eli's Drive-In (later part of the Ed Salem's chain) on Seventh Avenue South . A succession of popular disc jockeys, including Jim Lucas and Neal Miller, took turns spinning the most popular platters and delivering witty commentary and announcements. (Alvin Hudson collection.)

Bill Wright first gained attention for his *Merry-Go-Round 'Til Midnight* show on WSGN, playing records from 8:00 p.m. until midnight, but the children of the people who listened to him on radio knew him best as "Uncle Bill" on Channel 13. (Shaf Gregory collection.)

With nary a royal Canadian mounted policeman in sight, Shaf Gregory records an interview with movie singer Jeanette MacDonald. Shaf points out that he is using a wire recorder rather than tape, which was just coming into its own at the time. (Shaf Gregory collection.)

Radio personalities had to do just about anything to please their sponsors, and here WAPI's morning man, Leland Childs, is gamely demonstrating a chainsaw for one such advertiser. (Rosemary Lucas collection.)

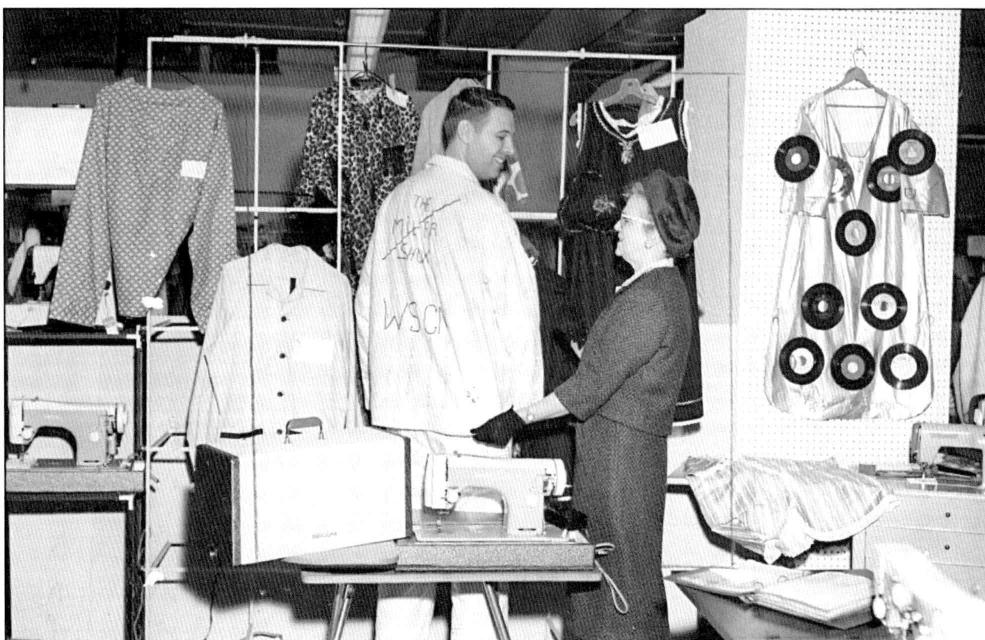

Because Neal Miller was so tall, WSGN conceived a promotional stunt in which listeners were invited to submit custom-made bathrobes to fit him. This unidentified lady was the winner; the robe in the background, with 45-rpm records sewn onto it, belonged to one of the losers. (Jean Spradlin-Miller collection.)

In July 1954, WBRC Radio switched its network affiliation from NBC to CBS; at the same time, WBRC-TV also made that change. (Author's collection.)

Station owner Dan Brennan was the guiding force behind WVOK's massive success during the 1950s and 1960s. (Patsy Preston collection.)

Baker Knight was a featured performer on WVOK in the 1950s. He later went on to a career as a nationally famous songwriter, penning tunes for such stars as Ricky Nelson and Elvis Presley. He passed away in October 2005. (Patsy Preston collection.)

Twelve-year-old Patsy Preston wowed Dan Brennan and WVOK audiences with her singing and guitar playing in 1957. She had been a fixture of WVOK's *Dixie Jamboree* shows since age eight. (Patsy Preston collection.)

WAPI was justifiably proud of its lineup of on-air talent in 1958. Many of them did double duty on television's Channel 13 as well, which would change its name to WAPI-TV during that year. (Rosemary Lucas collection.)

Future Birmingham radio legend Tommy Charles interviews some young ladies attending the showing of *Houseboat* at the Alabama Theater in 1958. (Alvin Hudson collection.)

In 1964, Doug Layton (shown here) and Tommy Charles left their stint at WYDE to hook up with radio station WAQY. The pair made national news with their publicity stunt of burning Beatles records after John Lennon was accused of saying the mop-topped quartet was "more popular than Jesus." (Author's collection.)

Although Davenport Smith admitted that he did not have the leading-man looks usually considered necessary for television success, he spent many years with the news department of WBRC Radio and was eminently respected by all. (Author's collection.)

Davenport Smith is using state-of-the-art tape equipment to record an interview with Gov. George Wallace, who is otherwise preoccupied with signing an autograph for an unidentified fan. (Author's collection.)

After beginning as WTNB, station WILD was occupying this building at the foot of Vulcan by 1954. It later featured a driveway and outdoor speakers so motorists could pull up to the window and request their favorite tune. (Alvin Hudson collection.)

WSGN had a float in the Zamora Shrine Circus parade through downtown on August 16, 1963. The significance of the kangaroo is unknown, unless WSGN was promoting marsupial madness because of the hit song "Tie Me Kangaroo Down, Sport." (Author's collection.)

Beautiful even with a mask covering most of her face, June West was the mysterious "Miss Midnight" on the all-night shift at WSGN. (June West collection.)

In the late 1950s, WSGN moved into this building on Seventh Avenue South and remained there until transferring to the penthouse of the City Federal Building in 1964. (Alvin Hudson collection.)

New on-air personality Ron Carney (far right) was welcomed to WAPI by, from left to right, staffers Hop Edwards, Cousin Cliff Holman, Dave Campbell, Dick McCullough, Jim Lucas, and assorted unidentified hostesses. (Cliff Holman collection.)

In the 1960s, WSGN was known for its group of announcers billed as "the Good Guys," who gave the listeners just what they wanted to hear and plenty of it. (Todd Jones collection.)

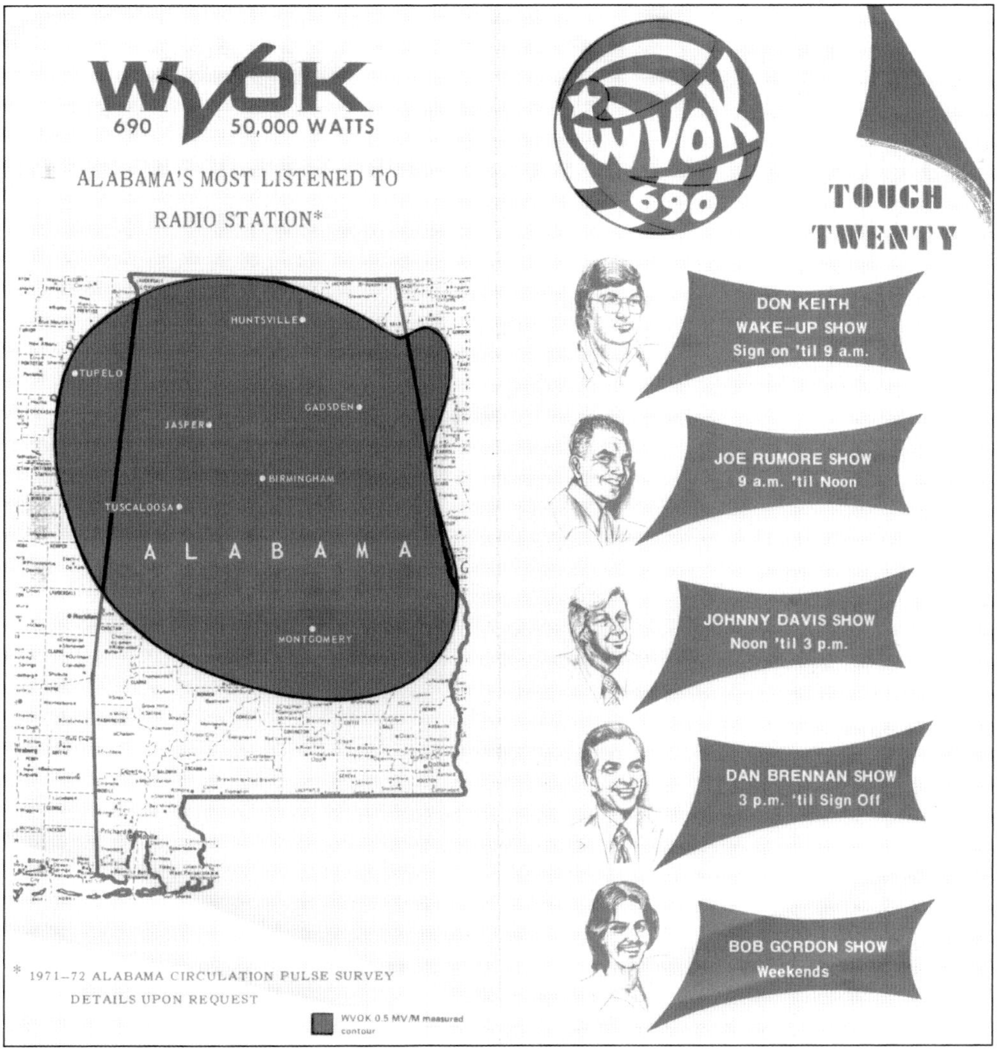

WVOK was still holding forth in 1975. The station would remain on the air until 1992; the 690 spot on the AM dial is now occupied by all-sports talk station WJOX. (Todd Jones collection.)

WSGN left the City Federal Building in 1978 and quit the airwaves forever on April 25, 1985. On July 28, 2005, just hours before demolition crews began their work of turning the City Federal Building into upscale loft apartments, the author rescued this vinyl decal off the wall of a small room on the building's main floor—the last evidence that WSGN ever operated out of that building. The WSGN call letters, incidentally, are now assigned to the public radio station in Gadsden, but with no connection to their original longtime Birmingham usage. (Author's collection.)

Two

Television Makes it Around That Corner

Television had been in the developmental stage for so many years that radio comedians took to making a running gag out of the expression, "Television is just around the corner!" Once World War II and its restrictions were over, the television industry was poised to take over the world of home entertainment, and the radio personalities were forced to laugh out of the other side of their faces, either going along with television or dropping out of show biz altogether.

Birmingham got its first two television stations within weeks of each other. Channel 13 was occupied by WAFM, the television branch of radio's WAPI. Channel 4 was the home of WBRC-TV, sister to long-established WBRC Radio and operated by the same person, Eloise Hannah Smith, who was certainly ahead of her time as a female broadcasting executive. Early in 1953, WBRC moved up the dial to Channel 6, because of some reception trouble in the northern borderline area, where the signal from Nashville's Channel 4, WSM-TV, could also be received.

The two Birmingham stations had received their licenses just before the FCC mandated a nationwide freeze on new stations, until some method could be figured out to handle the unexpected number of applications. One of those frozen applicants in Birmingham was WSGN, which fully intended to put WSGN-TV on the air as soon as the freeze was lifted. That never did happen. Later in the 1950s, WVOK wanted very badly to snag television's Channel 10, only to be told the number had been reserved for an educational station.

Of course, local programming was a big part of the Birmingham scene from the beginning. Channel 13 established itself early on with a number of locally originating shows, and WBRC went "live" in time for the 1952 presidential election returns. From then on, the two stations battled it out for ratings supremacy, even occasionally luring on-air talent from one to the other to tip the scale.

The FCC's solution that eventually ended the freeze was to open up the UHF band of channel numbers (everything over 13), and even though WSGN "fought like 610" to have a television version on Channel 42, it would be October 1965 before that channel became part of the Birmingham logs as WBMG. The reception for Channel 42 was so poor in most areas that it was frequently looked upon only as 6 and 13's poor unwashed stepchild.

Now let's turn on Channel 13 and see if we can't pick up a test pattern. Television has rounded that corner at last!

This test pattern began to be sent out over Channel 13 during the last week of May 1949. The station hit the air with its first programming on June 12, becoming Birmingham's first television outlet. (Author's collection.)

It is rather amusing to realize that there was no reason for anyone in Birmingham to buy a television set until there were stations on the air to be seen on it. Bromberg's was among the first stores to offer this newfangled technology to its customers. (Author's collection.)

Effective Sermon:

Horror Parade Chills

Deaths—

Richard H. Maxwell Dies; Burial To Be At Elmwood Cemetery

Funeral Services Held For Mrs. Susan Wise

Funeral Services Held For Seaboard Employe

Mrs. Robert E. Lawhon Passes In Atlanta

Rites Are Conducted For James Layfield

Miss Ruby Elizabeth Rich Passes At Home Here

Radio Log

5-HOUR SERVICE

CHAPMAN

Phonograph Records

SPECIAL . . 19c Each

6 for $1.00

BIRMINGHAM VENDING CO.

Admiral

PRESENTS ON

TELEVISION

TONIGHT and Every Friday Night!

"HOPALONG CASSIDY"

WILLIAM BOYD

WBRC-TV

CHANNEL 4

8 p.m. to 9 p.m.

Daily

TELEVISION

Demonstrations

12:30 and 6:45 P. M.

Radio Service Co.

Look Who LISTENS too!

MARK CHASE

"The Sheriff"

says—

"I get a real thrill listening to this kind of a stick-up — where people use gray matter instead of a gat to get the dough!"

Join Me in listening to

"BREAK THE BANK"

Tonight at

8:00 P. M.

WSGN

WBRC TECHNICIANS ON STRIKE

Radio technicians, Local 253, I. B. E. W., American Federation of Labor, advises that it has struck and is engaging in peaceful picketing against Radio Stations WBRC and WBRC-FM and TV.

The Union struck because of the unwarranted discharge of two Union technicians who had 20 and 8 years respectively with the station. These technicians had seniority over all other technicians. Four non-union technicians with less than one year's service were retained.

The company was charged with violating the Taft-Hartley Act. This charge has been withdrawn by the Union subject to refiling, if warranted by the evidence.

However, the charge referred to was not the basis of the strike, and the strike is continuing.

The Union respectfully requests your cooperation while this dispute is in progress!

C. M. BAKER, JR.

BUSINESS MANAGER

RADIO TECHNICIANS LOCAL 253

I. B. E. W., A. F. of L.

Tonight

ON

Screen Directors' Playhouse

CARY GRANT

IN

"Mr. Blandings Builds His Dream House"

8:00 P. M. CST

WBRC

Planning On Cosby?

Camp To Be Closed Over July Fourth

WAFM-TV

TELEVISION

ALABAMA

Here's Why

Channel 13

Is Your Lucky Number for

TELEVISION

Here Are the Famous Shows WAFM-TV Is Bringing Into Your Home

★ **Ed Sullivan's Toast of the Town**

Sundays, 8:00 P.M.

Tonight

6:30 P.M.

"An Evening of Starlight Opera"

Come Out and See WAFM-TV Cameras in Action!

★ **It Pays to Be Ignorant**

Mondays, 8:00 P.M.

★ **Mr. I. Magination**

Sundays, 7:30 P.M.

★ **54th Street Review**

Thursdays, 7:30 P.M.

★ **Where's Wesley?**

Sundays, 7:00 P.M.

★ **John Wayne Westerns**

Saturdays, 7:00 P.M.

★ **Blind Date**

Wednesdays, 7:30 P.M.

★ **Hollywood Screen Test**

Mondays, 7:30 P.M.

★ **Television Theater**

Fridays, 7:30 P.M.

★ **Lucky Pup**

Mondays thru Fridays, 7:00 P.M.

★ **Wrestling**

Mondays, 9:00 P.M.

Daily

12-2 P.M.

Channel 13

Test Pattern and Programs

★ **Baseball**

Channel 13 . . .

WAFM-TV

"Television Alabama"

July 1, 1949, was a big day in Birmingham, as that was when WBRC-TV finally made it on the air, and Birmingham had two stations from which to choose. This is the entire radio/television page for that day, complete with an announcement that WBRC's technicians were already on strike. (Author's collection.)

For the first year or so, Birmingham could not see network television shows when they were originally broadcast, instead relying on filmed (and thus delayed) programming. On September 30, 1950, Birmingham was finally joined to the coaxial cable that made it possible for everything to be seen at their proper times. (Author's collection.)

Having already encouraged people to buy television sets in the first place, Bromberg's doubled its efforts once network television was the big deal in town. There was still the difficulty in dividing four networks' offerings—NBC, CBS, ABC, and DuMont—among only two stations. (Author's collection.)

As operations manager, Lionel Baxter was one of the guiding forces behind the early days of Channel 13. He later moved his influence to Channel 6, and from there to the Philadelphia television market. (Alvin Hudson collection.)

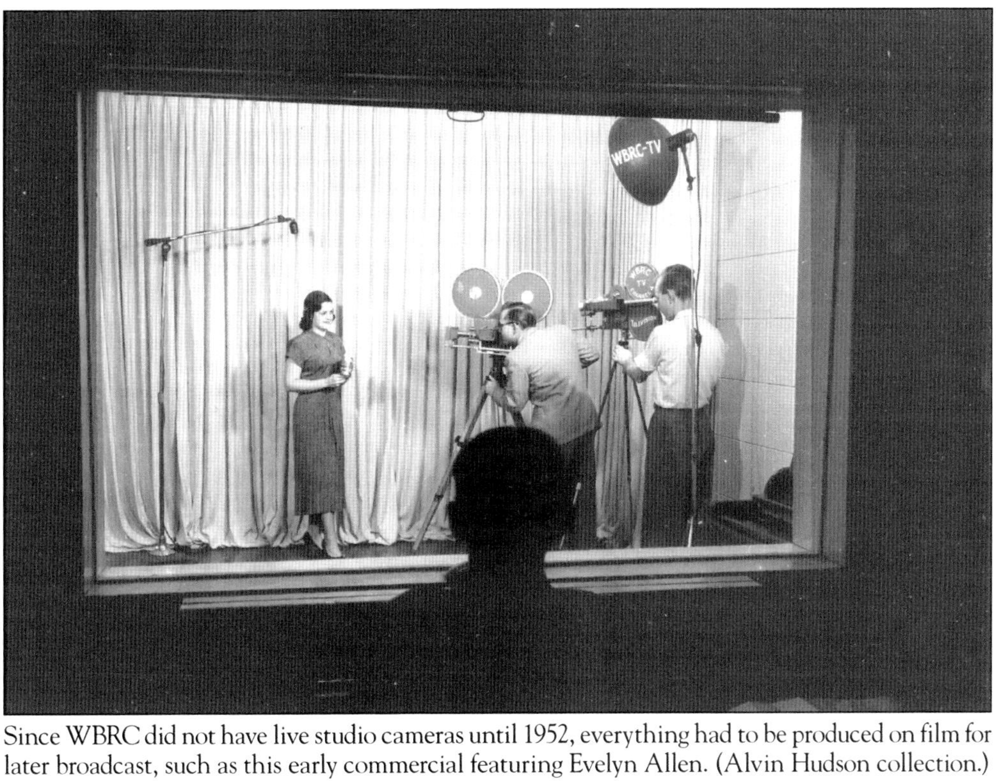

Since WBRC did not have live studio cameras until 1952, everything had to be produced on film for later broadcast, such as this early commercial featuring Evelyn Allen. (Alvin Hudson collection.)

Election night, November 4, 1952, was a memorable day at WBRC, as the newly installed studio cameras enabled the station to broadcast the returns live. Notice from the signs on the wall that WBRC was still known as Channel 4 at this time; it would become Channel 6 early the following year. Not everyone in this photograph has been identified, but the fellow standing in front of the counter is newscaster Alvin Davis, and seated immediately next to him is sales-whiz Keith Williams. At the far right are Annette Griffis (later of WSGN) and Shaf Gregory. (Alvin Hudson collection.)

Channel 13 used this converted bus to transport equipment and personnel for remote broadcasts during the 1950s. (Alvin Hudson collection.)

After beginning life as WAFM, by 1953, Channel 13 had become WABT (for Alabama's Best Television), with the creative logo seen on these cameras. (Alvin Hudson collection.)

The first big network shake-up in Birmingham television came on July 4, 1954, when NBC moved from Channel 6 to Channel 13. The network remains with that station today, but Channel 6 would see several more changes over the next few decades. (Author's collection.)

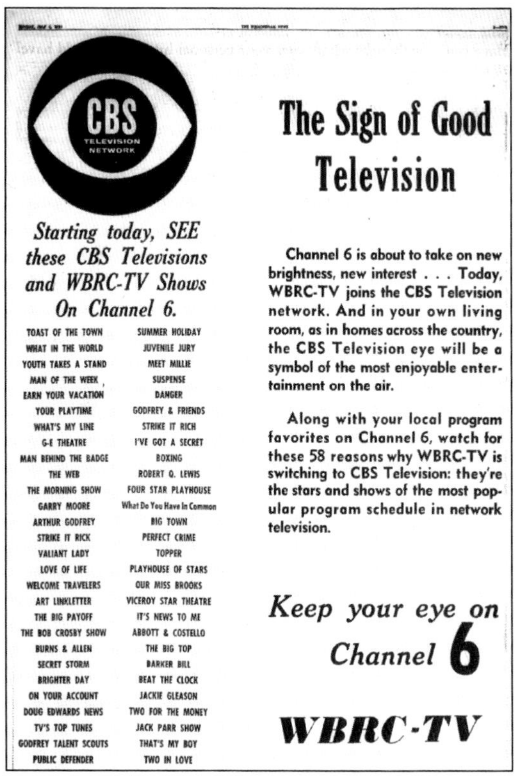

At the same time NBC moved to Channel 13, CBS set its eye on Channel 6. It would remain with WBRC until 1961. (Author's collection.)

WBRC's purchase by television giant Storer Broadcasting brought it new power and influence, as seen in this trade industry ad from the summer of 1954. (Shaf Gregory collection.)

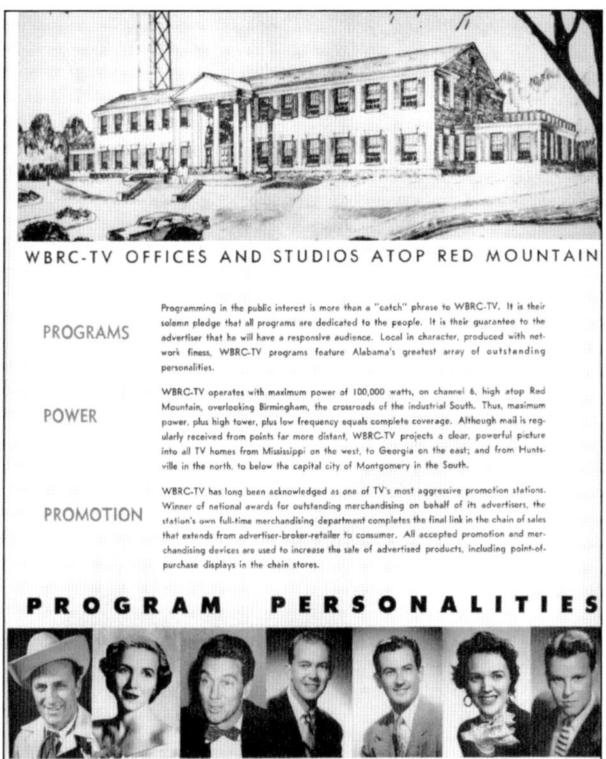

Storer used the inside of its flyer to promote WBRC's forthcoming new studios and its popular programming personalities including, from left to right, Happy Hal Burns, Joan Meadows, Rick Nelson, Bob Bandy, Henry Norton, Margo George, and Shaf Gregory. (Shaf Gregory collection.)

In September 1954, WBRC threw open the doors of this imposing Southern plantation-style studio building directly in front of the old WBRC-FM structure. (Alvin Hudson collection.)

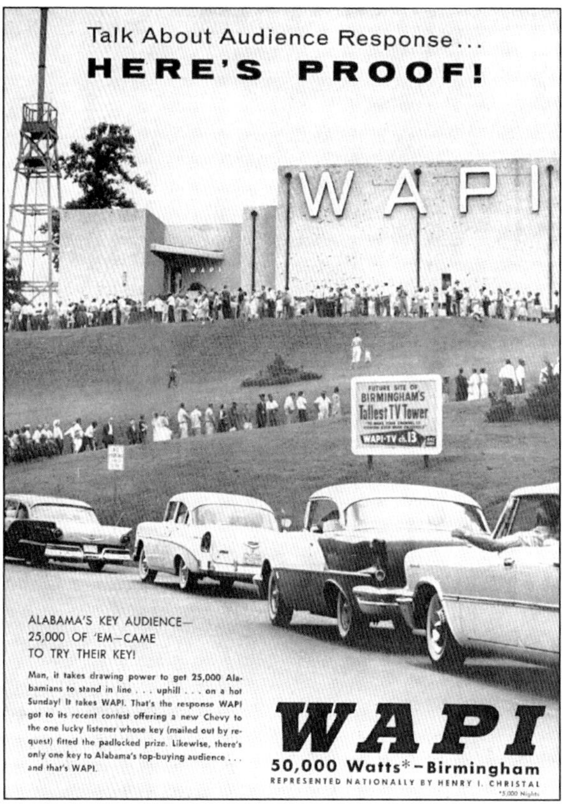

While WBRC was building a new studio, WAPI Radio and Channel 13 moved into the former WSGN facility on Red Mountain. In 1959, this crowd lined up in front of the station for a chance to win a new car. (Rosemary Lucas collection.)

This February 1962 view of Red Mountain looking south shows Channel 13's dual buildings, which would soon be combined into a single structure. (Alvin Hudson collection.)

Everett Holle served as Channel 13's program director for four decades. He also stepped before the cameras to do various on-air work, including emceeing and interviews. (Everett Holle collection.)

By 1961, it was time to switch television networks again. CBS left WBRC and shared time with NBC on Channel 13. At a kickoff ceremony, Everett Holle is welcoming one of CBS's current stars, Bob Denver of *The Many Loves of Dobie Gillis*. (Author's collection.)

Meanwhile in March 1961, Channel 6 became an ABC affiliate. Peter Brown (center) of *The Lawman* helped Pat Gray and Tom York indoctrinate viewers to the station's new network lineup. (Bill Bolen collection.)

WBRC's Fannie Flagg helped with the station's all-out push for the new 1964 ABC-TV season. (Fannie Flagg collection.)

Harry Mabry (left) moderated a press conference consisting of, from left to right, ABC network stars Paul Petersen, Ted Hartley, Peter Jennings, David Hedison, and Ken Berry. (Author's collection.)

To publicize ABC's impending sitcom smash *Bewitched*, Fannie Flagg appeared in comical promotional spots as an old hag known as "Natalie Natfinger." (Author's collection.)

Pat Gray (left) devotes some WBRC airtime to an interview with Eileen O'Neill of ABC-TV's *Burke's Law*. Somehow most of Channel 6's male personalities managed to have their photographs made with O'Neill during her visit too. (Author's collection.)

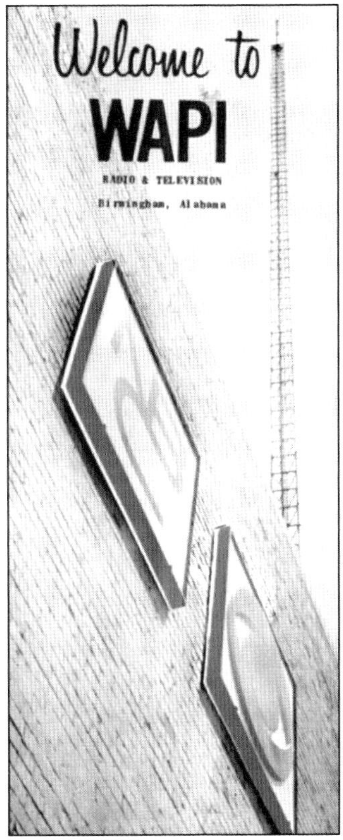

Channel 13's dual affiliation with both NBC and CBS from 1961 to 1970 gave viewers fits when trying to figure out when network programs could be seen. Many shows from either network were seen on a delayed basis or not at all. (Wendell Harris collection.)

When WBMG Channel 42 went on the air in October 1965, it initially picked up the leftover NBC and CBS shows not aired by Channel 13. In 1970, WBMG became a full CBS affiliate and left the NBC shows with WAPI. (Author's collection.)

A historic event for Birmingham television: longtime WBMG general manager Hugh Smith (left) and station owner Bill DuBois (right) watch CBS-TV executive Carl Ward sign the contract making Channel 42 a full-time CBS affiliate. For the first time since 1949, Birmingham had full affiliates for all three television networks. (Hugh Smith collection.)

Stations often took out special holiday ads, such as these that appeared in *TV Guide* during the 1970s. (Author's collection.)

Independence Day was a big deal at WBRC, with speeches from military personnel and politicos galore. Colonial minutemen are running the flag up the pole to see if anyone salutes. (Author's collection.)

A 21-gun salute in WBRC's parking lot also gives us a good view of the rarely-noticed second neon sign that overlooked Homewood. Its twin on the downtown side of Red Mountain became much more of a local landmark. (Author's collection.)

Channel 13's face wasn't red in April 1967 when it finally got color cameras in the studio, allowing all of its local programming to be broadcast in color for the first time. (Author's collection.)

The cerebral palsy telethon was WBRC's primary charitable project each year. In 1964, Fannie Flagg (left) and Pat Gray don the uniforms of the cosponsoring U.S. Marines to promote the cause. (Pat Gray collection.)

At a later CP telethon, Pat Gray is flanked by, from left to right, Peter Jennings of ABC News, David Hedison of *Voyage to the Bottom of the Sea*, Paul Petersen of *The Donna Reed Show*, and Ken Berry of *F Troop*. (Pat Gray collection.)

After being sold to Times-Mirror Broadcasting in 1980, Channel 13 gained the fourth set of call letters in its history, WVTN, shorthand for Vulcan Times-Mirror. Will they be its last? Only Times-Mirror will tell! (Ernest Langner collection.)

In a symbolic passing of the torch, Hugh Smith (right) turned the reins of WBMG general manager over to comparative youngster Hoyle Broome in 1981. Channel 42 would become known as WIAT (representing its new slogan, "It's About Time") in 1998. (Hugh Smith collection.)

Three

And Now for the News

From the very beginning, news was intended to be one of the most important parts of television, just as it had been in radio. The ability to report on important events while they were taking place was something that the daily newspaper could not hope to equal, but in the early days, the primitive television news-gathering equipment didn't make the job much easier.

It was possible to broadcast live from a remote location outside the studio, but it was rarely done because of the trouble and expense involved. Most footage was shot silent for the news anchor's explanation to be heard over it as it ran.

Mostly coming from radio, those great old news anchors were all stentorian in their tone and unflaggingly serious on the air, no matter what sort of wacky senses of humor they might have possessed off-screen. Joe Langston, one of Birmingham's most respected newscasters, has said that a favorite trick of the floor crew was to make the anchor crack up during the broadcast, a feat they would try to accomplish by pasting a Playboy centerfold in the place of the regular Teleprompter script and other such antics.

Then there was the unique breed of television personality known as the "weather girl." Although it would be considered sexist in the extreme today, the fact was that the weather girl's main function was to provide an eye-soothing alternative to the serious newsmen.

Dick Breit was one of the few non-females to hold down the weather position on the news set. To Dick, the bigger picture was that, for every news or entertainment personality, television viewers fell into three categories. "The first group was the people who just liked you for no particular reason," he says. "No matter how good or bad you were, they were loyal to you. The next group was people who absolutely couldn't stand you, again for no particular reason other than your personality rubbed them the wrong way. Then, there was the third group. Those were the people who, if you asked them what they thought of Dick Breit, would answer, 'Who?' "

And that about sums it up. Whether they were known for being utterly professional or utterly obnoxious, whether they sported supermodel good looks or wore a toupee, whether they had been in radio for decades or were just starting out in the business, these were the people we depended on to tell us what was happening beyond the confines of our home's four walls. Let's see what's in the news today.

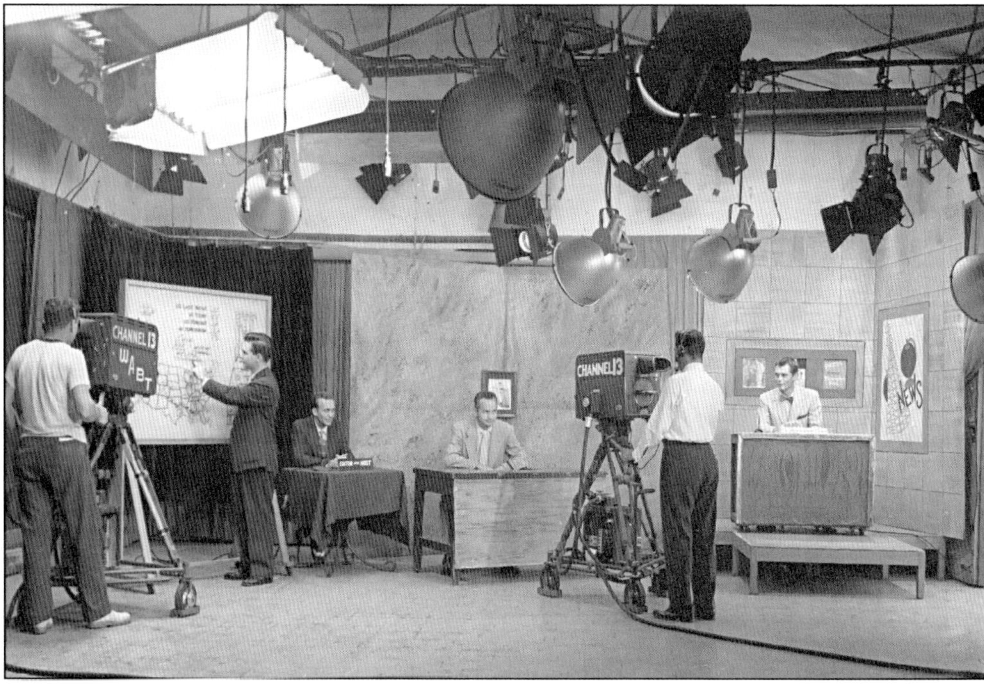

"Slick" and "sophisticated" were not two words that described local television news in the early days. Channel 13's newscast in 1953 included, from left to right, Bart Darby at the weatherboard, Charlie Davis, Dan Daniel, and Sterling Brewer with sports. (Alvin Hudson collection.)

Meanwhile over at WBRC, Shaf Gregory was delivering news gems for Busch's Jewelers. (Shaf Gregory collection.)

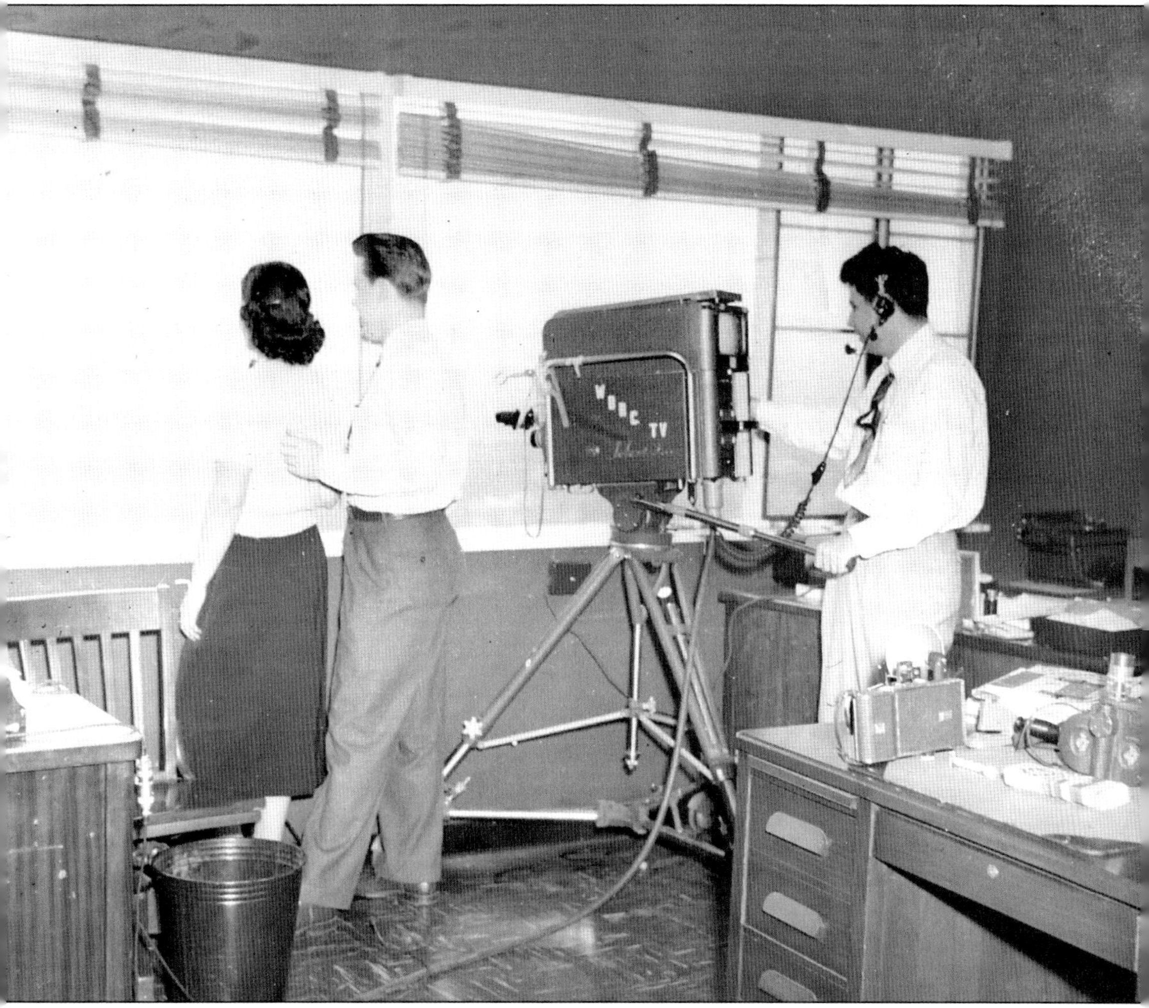

Speaking of sophisticated news-gathering methods, Shaf Gregory (center) explained what was going on in this photograph. There was a fire at Five Points South, and WBRC needed to broadcast the image from its perch high atop Red Mountain. Solution: they got a pair of cheap binoculars from Woolworth's, taped them to the camera lens, and according to Shaf, the resulting television picture made it appear that a camera crew was actually on the scene. Annette Griffis is helping Shaf keep an eye on the flames while cameraman Jack Hand handles the binoculars. (Shaf Gregory collection.)

After a stint at Channel 13 in the early 1950s, Harry Mabry joined the WBRC news department and soon became a television legend. (Bill Bolen collection.)

Dick Breit forecast the weather on Channel 13 for some 20 years or more. He made news headlines of his own in 1983 when he abruptly decided to stop wearing his toupee for his newscasts. (Dick Breit collection.)

Dick Breit was not the only one who had hair trouble at Channel 13. When Rosemary Lucas joined the staff, the crew could not figure out how to properly light her jet-black hair, so they made her dye it blonde for a short time. (Rosemary Lucas collection.)

Fortunately the technicians got it all figured out, and she was able to become the Rosemary we all know and love. (Rosemary Lucas collection.)

Bart Darby was a member of the WBRC news team in the early 1960s, but some people remember him best as the station's original Bozo the Clown. (Bart Darby collection.)

In spite of his youthful appearance, Joe Langston became one of Birmingham's most respected and trusted newscasters after joining WBRC in 1963. (Joe Langston collection.)

Do you remember when television forecasters were known as "weather girls" instead of meteorologists? The equipment and skills needed have matured, but people still basically want to know whether or not it's going to rain tomorrow. (Rosemary Lucas collection.)

Channel 42 bucked the "weather girl" trend with its forecasts by Bob Litton. No one recalls who the guest lady in this shot could have been, but at far right is part of a sign plugging the newscast's sponsor, Jefferson Federal Savings. (Bill Wilson collection.)

Harry Mabry (left) and Tom York (center) handled the local news on WBRC in the mid-1960s. Joe Langston, as befitting his global backdrop, stood by with national news. (Bill Bolen collection.)

The Channel 13 gang in the late 1960s included, from left to right, anchor Wendell Harris, sports reporter Gary Sanders, "weather girl" Rosemary, and news director Bob Jones. (Rosemary Lucas collection.)

After serving as sidekick on *The Morning Show*, Pat Gray moved up to Channel 6 "weather girl" in 1961. Her portion of the newscast was sponsored by Lorch's Jewelers, and Harry Mabry always introduced her as "Lorch's most precious gem." Few viewers disagreed. (Pat Gray collection.)

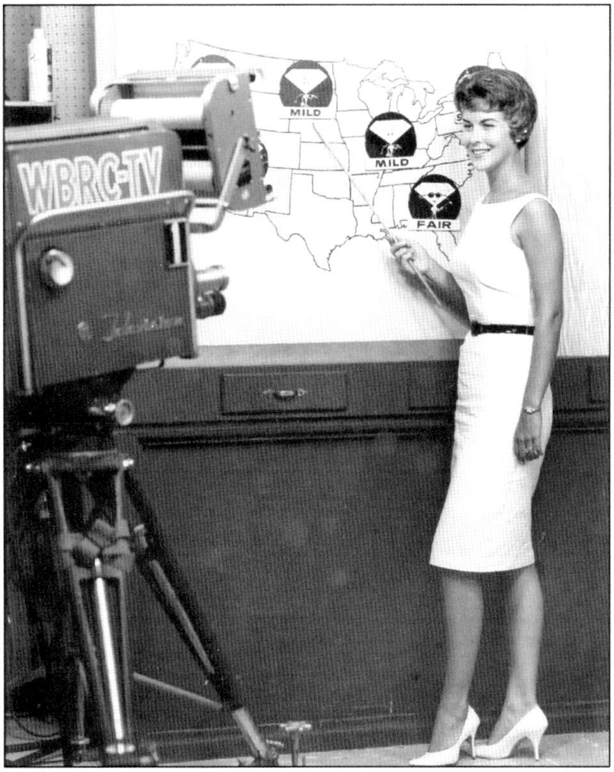

Whereas Rosemary had earlier switched from blonde to dark hair, during the 1960s Pat Gray moved from her short brunette hairstyle to this blond flip that became her trademark. (Author's collection.)

Larry Langford eventually became a big wig in local politics, but as a WBRC news reporter in the 1970s, this is not a big wig he is wearing here. (Author's collection.)

Behind Wendell Harris's stern and solemn on-air demeanor lurked a sense of humor that viewers at home rarely, if ever, got to see. (Wendell Harris collection.)

Radio fixture Tommy Charles gamely hosted the sports segment of Channel 42's newscast for a time, but reports are that he treated it much as the joke everyone else considered it to be. (Author's collection.)

Gary Sanders and Wendell Harris calmly discuss the traditional Alabama-Auburn rivalry during a 1970s sports program. (Wendell Harris collection.)

The debut of the "NewsWatch" format in 1975 simply meant that Channel 13 stood by with a full-time news team ready to break in with a developing story at any moment. A few years later, the "NewsWatch" format was replaced by the "Action News" format, which Wendell Harris summarizes as meaning, "If it bled, it led." (Author's collection.)

When it came to television news, WBMG was the Rodney Dangerfield of Birmingham broadcasting. No matter how many attempts were made to compete with Channels 6 and 13, Channel 42's UHF signal and Johnny-come-lately status always worked against it. News anchors moved in and out as if the station had a revolving door. John McGiboney (above), John Dawson (below), and Bob Harper (bottom) all rotated holding down the news desk during the period of a year. (Author's collection.)

Wendell Harris and Rosemary pose with Channel 13's state-of-the-art news van, *c.* 1978. This was a far cry from the primitive bus we saw back on page 41. (Rosemary Lucas collection.)

Herb Winches became the sports anchor most identified with WBRC in the 1970s. In 2006, he could still be heard on all-sports radio station WJOX, the former WVOK. (Tom York collection.)

This advertisement from April 1976 wasn't just publicity hype. WBRC's newscast was consistently a ratings leader, and its anchors were local superstars. Of course, in those pre-cable television days, the fact that Channel 6 had better picture quality than either of its competitors didn't hurt matters any. (Author's collection.)

For one brief period, from 1980 to 1982, WBRC's star anchor, Joe Langston (left), jumped ship and appeared on Channel 13's newscast with, from left to right, James Spann, Pam Huff, and fellow WBRC defectee Herb Winches (sans mustache). Both returned to Channel 6 afterward, and all was once again right with the world. (Joe Langston collection.)

In today's Birmingham television news market, Bill Bolen (left) is the veteran after more than 50 years behind the mike. Here in 1982, he is accompanied by Colleen Bates and Tom York. Colleen must be sitting down, because in real life she towered over both of her costars. (Tom York collection.)

Four

Cowboys, Captains, Clowns, and Cartoons

The news might have been well and fine for stodgy old parents and other alleged adults, but remember that the 1950s and 1960s weren't called the "baby boom" for no reason. During those years, children increasingly made up the largest part of the country's population, and the mass media were not slow to respond. Network television was quick to see a potential audience in those youngsters, and local stations across the nation soon followed.

In Birmingham, it all started with Benny Carle. WAFM had purchased a batch of old Western movies staring the likes of Hoot Gibson and Tom Mix, and Benny was faced with the challenge of filling in the time not occupied by the commercial breaks in the chopped-up flicks. He soon hit on the idea of inviting groups of kids to sit in the studio with him. In the allotted time, he would interview the young buckaroos about their names, schools, hobbies, and so forth. "At first we got mostly Scout troops, because they were already organized and easier to get to the station," Benny says today. "Later, birthday parties took over, because moms could plan those, too."

Channel 6 did not try to counter-program *Western Theater* at first, mainly because it had no live cameras in its studio. That changed in October 1952, and WBRC soon presented WSGN announcer Horace Pumphrey as Supersonic Sam. After that came Louis Sudduth as Mister Magic, and finally Bob Bandy as the host of *Circle Six Ranch*, an unabashed clone of 13's program. When all of these flopped, WBRC did the smart thing and offered Benny Carle enough greenbacks to entice him to hitch up his cowboy britches and mosey over to the competition. The battle lines were then drawn, and the ratings war began in earnest.

The death of local kids' programming in the late 1960s and early 1970s was a nationwide phenomenon brought about by the stations's discovery that syndicated reruns could attract just as many viewers as hosted shows, with a much smaller outlay of capital. The changing demographics that resulted from the maturing of the Baby Boomers were also a factor. For a complete history of the phenomenon that was local kids' television, consult the book *Hi There, Boys & Girls! America's Local Children's TV Programs* (University Press of Mississippi, 2001). But for now, grab a Twinkie and a glass of Consolidated Dairies milk, and get ready to be entertained by some truly great entertainers.

Wearing a cowboy shirt borrowed from Happy Wilson, Benny Carle began showing vintage oaters on Channel 13's *Western Theater* only weeks after the station went on the air in 1949. (Benny Carle collection.)

Western Theater was the first kids' show in Birmingham to feature a studio audience of children. Benny found that he had a natural rapport with the younguns. (Benny Carle collection.)

During the 1950 Christmas season, Cliff Holman got his first experience with a daily television show when he was hired as the puppeteer for *Mr. Bingle*, sponsored by Loveman's and broadcast from the third floor of the department store. (Cliff Holman collection.)

When Benny Carle left Channel 13 in 1954 to take his show to Channel 6, his replacement was "Uncle" Bill Wright, who had previous kid-show experience as sidekick to Miss Ann on WSGN's *Comic Club*. The junior Easter bunny is Bill's daughter Kelly. (Bill Wright collection.)

In March 1954, the Ward Baking Company installed Cliff Holman in a new daily 15-minute series, *The Tip-Top Clubhouse*, and gave their youthful star a new nickname. He would be "Cousin Cliff" for the rest of his career. This publicity photograph for *Tip-Top Clubhouse* features Cliff with his costars, Kim and Corky. (Alvin Hudson collection.)

Uncle Bill Wright's trademark was his antique car, which was named "Bouncalot" in a station contest. When Uncle Bill followed Benny Carle's lead and defected to WBRC in 1956, Cousin Cliff inherited both his show and Bouncalot. (Cliff Holman collection.)

Not all kid shows had the staying power of Benny Carle or Cousin Cliff. Newscaster Sterling Brewer's *Balloon Goon* show aired in 1957 before deflating like a tired balloon. (Cliff Holman collection.)

The highly successful *Romper Room* franchise, based out of Baltimore, made its Birmingham debut on WBRC in October 1955. "Miss Jean" Prescott was the first schoolmarm to keep things lively. (Shaf Gregory collection.)

Young Martha Cohen spent some quality time with Miss Jean in September 1956. The Channel 6 *Romper Room* lasted only two and a half years, but the format would return on Channel 13 during the following decade. (Martha Cohen collection.)

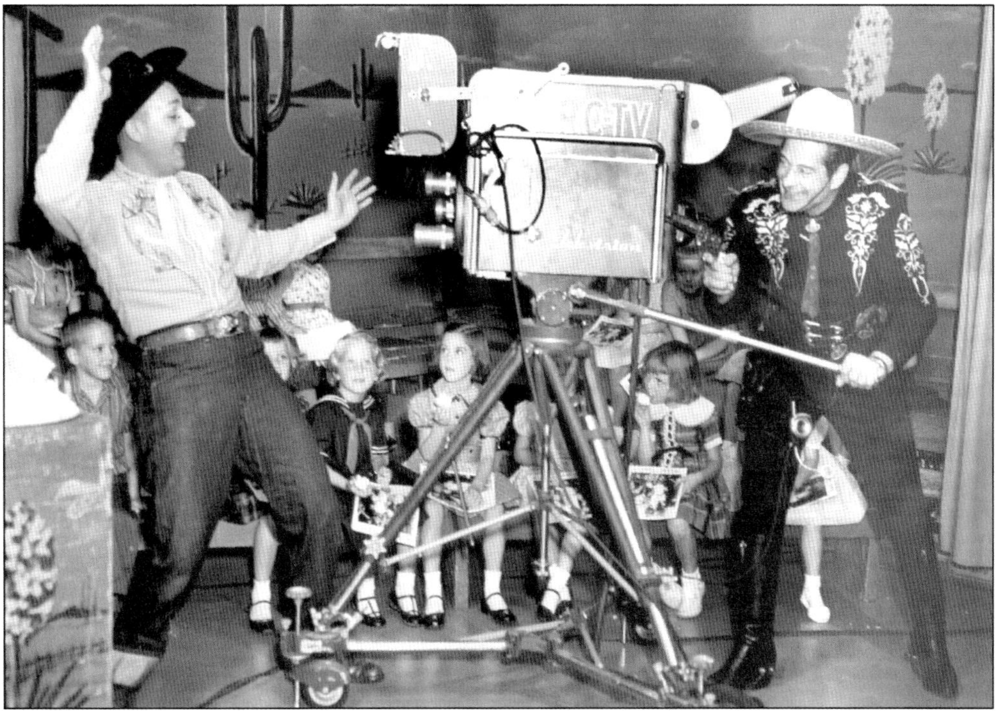

Benny Carle continued his established Western theme on WBRC with *Circle Six Ranch*. Here he engages in a bit of tomfoolery with Duncan Renaldo, alias the Cisco Kid. (Benny Carle collection.)

Benny appears to be a bit stumped by the amount of his fan mail. Perhaps he is enlisting the burro's help in reading it all. (Benny Carle collection.)

Bozo the Clown was another successful national franchise that made its way to WBRC. Although initially reluctant to do so, newsman Bart Darby donned the baggy blue suit and heavy makeup to host the show each weekday. (Bart Darby collection.)

In 1958, Cousin Cliff began hosting the wildly popular Popeye cartoons, and the one-eyed spinach eater would be his close buddy from then on. "Popeye bought me two houses and four cars," says Cliff. (Greg Glenn collection.)

Western movies could no longer compete with Popeye, so WBRC counter-programmed by making Benny Carle the host for the Bugs Bunny cartoon package. In this early shot, Bugs and guest star Edward Everett Horton help Benny suit up for his new format. (Benny Carle collection.)

Not all local kid shows were wacky and zany. Pat Gray won much recognition for her *Young People's World*, which boiled the daily news down to a format children could comprehend. (Pat Gray collection.)

WBRC general manager Bob Schlinkert (right) goes over a *Young People's World* script with Pat Gray and series producer Charlie Metcalfe. (Author's collection.)

Cousin Cliff's *Popeye Show* became the hottest ticket in town for kids who wished to celebrate their birthdays on television. This was his most famous set, painted by Cliff's father, who was also Channel 13's art director. (Cliff Holman collection.)

Before leaving Birmingham television in 1964 to take his act to Anniston and Huntsville, Benny Carle was also conscripted to preside over *Junior Auction*, on which kids used empty Golden Flake potato chip bags to bid on toys and other goodies. (Alvin Hudson collection.)

After Bart Darby departed WBRC for New Orleans in 1962, his dual jobs as newscaster and Bozo were assumed by ex-WSGN personality Ward McIntyre, who clowned around until the show ended in 1968. (Ward McIntyre collection.)

The kid show hosts got out of their studios to make many personal appearances such as this one, where Ward "Bozo" McIntyre is getting acquainted with two adorable young fans at the Children's Hospital. (Ward McIntyre collection.)

Channel 6 sent out these glossy color postcards to kids who wrote in to their favorite clown. Anyone who took on the role of Bozo had some mighty big shoes to fill. (Author's collection.)

While Channel 13 was carrying the syndicated reruns of *The Mickey Mouse Club* in 1964, they quacked everyone up with an appearance by Clarence Nash, the original voice of Donald Duck, at Eastwood Mall. (Author's collection.)

On his *Bugs Bunny Show*, Benny Carle tuned in Looney Tunes on a fake television set he called the Framischlocker. When asked why he called it that, Benny always replied, "Because it looks like a Framischlocker." (Benny Carle collection.)

When Channel 42 debuted in October 1965, WSGN's Neal Miller stepped in front of the camera to host Dick Tracy cartoons as "Sergeant Jack." He was, of course, named after his primary sponsor, Jack's Hamburgers. (Jean Spradlin-Miller collection.)

In order to wear the authentic Jefferson County Sheriff's Department uniform, Neal Miller had to be officially sworn in as a deputy. He always laughed that people didn't realize he could have legally arrested someone if the situation came up. (Jean Spradlin-Miller collection.)

Channel 13 revived *Romper Room* in April 1966 with "Miss Jane" Hooper behind the desk. When kids started to school, they were no doubt disappointed when their real teachers were not as attractive as Miss Jane. (Author's collection.)

Story time and snack time were a good fit on *Romper Room*. The Dortch bakery kept the kids well supplied with its line of Kim and Jim snack cakes. (Jane Hooper collection.)

At the end of each show, Miss Jane violated who knows what sort of privacy laws by using her Magic Mirror to look in on the kids at home. "I see Judy and Bobby and Suzie and . . . " (Author's collection.)

Family commitments forced Miss Jane to leave *Romper Room* in 1968, and the new teacher was petite and perky "Miss Carol" Aldy. When the show moved to Channel 42 for its final year (1970–1971), Miss Carol went along too. (Carol Aldy collection.)

Because Miss Carol was seen only with the preschoolers on the show, when making personal appearances she often surprised viewers when they saw she was barely over five feet tall. (Carol Aldy collection.)

Although he was better known as a news anchor, Joe Langston was assigned to a Saturday morning children's show known as *Birthday Party* from 1966 to 1968. The disembodied hand in the contest box, known as "SomeThing," belonged to the show's floor manager, who later gained far more notoriety as "Slamming Judge" Jack Montgomery. (Joe Langston collection.)

In his newly designed, groovy, psychedelic set laced with flower power, Neal Miller interviews child actors Johnny Whittaker of *Family Affair* and Brandon Cruz of the *Courtship of Eddie's Father*. The *Sgt. Jack Show* aired on weekdays until September 1976 and was seen sporadically on weekends until 1982. (Jean Spradlin-Miller collection.)

The first puppet character to interact with Sergeant Jack was the colossal con man Wilbur the Groundhog, performed by WBMG art director Howard Cruse. (Howard Cruse collection.)

Ted Lowry's puppets were added to the *Sgt. Jack Show* cast in the early 1970s. When Jack's dropped its sponsorship of all kid shows in 1970 and McDonald's stepped in, it made for an odd juxtaposition of Sergeant Jack's name with his new sponsor. (Jean Spradlin-Miller collection.)

Five

A Word from Our Sponsor

Let's face it—none of the shows we have reminisced about so far could have lasted for very long if it had not been for those all-important commercials that helped pay the bills. When radio first began to show evidence that it could be a powerful mass communication tool, the debate immediately began as to whether it should remain free of monetary influence and seek the high ground, or whether the airwaves would be sold to whoever had the necessary funds to pay for the privilege. We all know which side won that argument.

Birmingham's radio and television advertising became such an integral part of the overall broadcasting scene that many of its sights and sounds are as well remembered as the entertainment or news shows they interrupted. What about the decades-spanning radio jingle for Buford Bryant, "known as the Tire Giant?" His commercials proved without a doubt that "You'll desire General Tires, 'cause Bryant never tires of serving you." Who could resist singing along to the signature cha-cha-cha tune for Bargain Town U.S.A., where "They buy for less so they can sell for less; the money you save will bring you happiness?"

Once a visual image was added to the catchy slogans, the commercials became even more maddeningly unforgettable. How many times per day did we see Harry Charles trying to sell us a piano by reminding us, "To pay more is sheer extravagance?" If it wasn't Harry, there was Ed Phipps, who reminded us of his proximity to giant music store E. E. Forbes by hinting, "Walk a block and save." Taft Epstein and his Epp's Jewelry came on the air to the incessant pounding of his wind-up toy monkey. Department stores lured customers with elegant tunes such as "We wrap a smile with every gift from Parisian," and "You've got style, you've got Pizitz." Ford dealer Jim Skinner (and later his two sons) pulled off the corniest gags through sheer bravado.

There is little doubt that the Birmingham company that made the best consistent use of television advertising was Jack's Hamburgers. Former Pig Trail Inn restaurateur Jack Caddell chose to sink 100% of his advertising budget into the Birmingham children's shows. The chain went even further when WBMG arrived, naming its new host Sergeant Jack.

So, don't get up to get a snack or go to the bathroom while the commercials are on, or you'll miss the entire next section.

Just because WSGN was an ABC affiliate, that didn't stop them from advertising National Baking Company bread. Announcer Jim Cleary seems to be trying to break a record by proving that the bread makes a delicious sandwich out of anything. (Shaf Gregory collection.)

Mother's Best flour was a perfect fit for Joe Rumore's homespun style on WVOK. (Author's collection.)

Happy Hal Burns's program featured a sped-up voice he referred to as "Gismo." In February 1949, WBRC apparently had a contest for listeners to submit their ideas of Gismo's physical appearance, and this unidentified lady was the winner. At right, helping Happy Hal present the prize, is automotive dealer Jim Burke. The poem in Gismo's left hand reads, "Hi, folks! I'm Gismo of the Happy Hal Burns Show / I'm fat and round and not too high from the ground / My jokes and songs over WBRC fill every heart with glee / I'm quite a fellow, full of pepper and dash, in my car from Jim Burke Nash." (Alvin Hudson collection.)

Happy Hal Burns and sidekick Gismo continued their happy association with Jim Burke Nash during this 1949 promotion shared with Gold Medal Flour. (Alvin Hudson collection.)

Automobile dealer Jim Skinner (right) became famous for his hilariously hokey television commercials, but at this time he was a representative for O. Z. Hall Ford. He and fellow salesman John Israel (left) are heartily congratulating Dan Brennan on WVOK's sales job for the dealership. (Patsy Preston collection.)

As one of Birmingham's largest retailers, Pizitz made good use of television over the decades. This advertisement for the store's popular *Charades* show appeared in 1950. (Author's collection.)

Twenty-two years later, in 1972, Pizitz was still on television, using its commercial time in this Thanksgiving cartoon to promote the annual Enchanted Forest display on the sixth floor. (Author's collection.)

Mimi Dennis handled women's affairs for WAPI and took time out to promote Sofskin hand cream. (Alvin Hudson collection.)

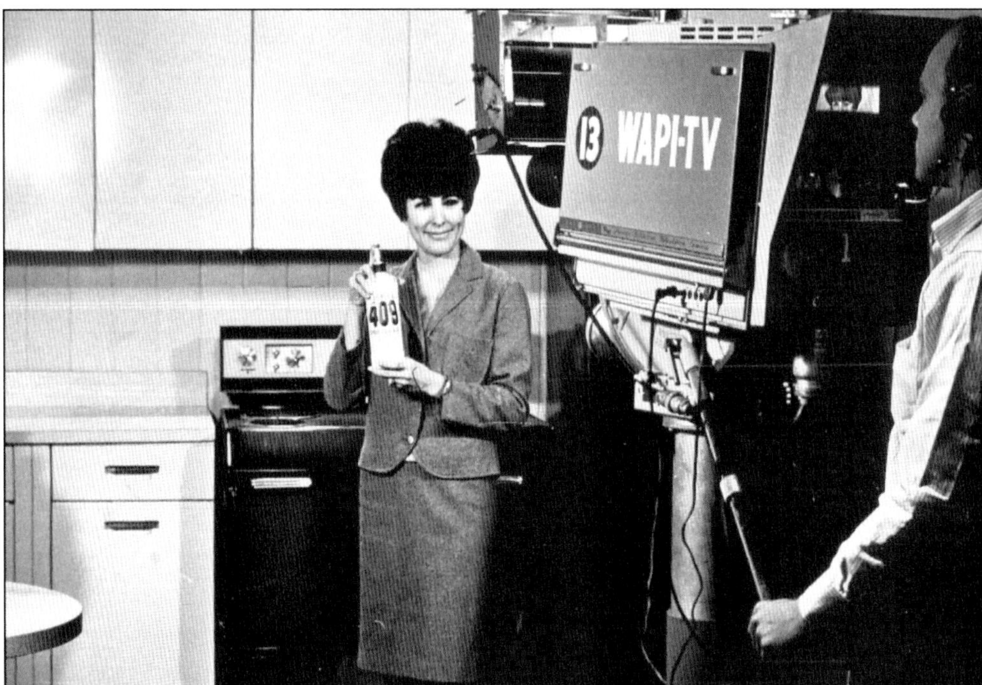

Rosemary was more likely to be found behind a mic than in the kitchen, but she used her *Midday* show to plug that new miracle cleaner, 409. (Rosemary Lucas collection.)

Yellow Label syrup was an aggressive radio advertiser for decades, and obviously had no trouble when it came time to sop up sales in television. They must have helped Hopalong Cassidy out of some pretty sticky situations. (Shaf Gregory collection.)

No one is certain why Jim Lucas was wearing this sort of outfit to promote Birmingham's native Buffalo Rock Ginger Ale. (Rosemary Lucas collection.)

Birmingham had an amazing number of jewelry stores for a city its size, and Busch's was right in there mixing it up with the rest. Shaf Gregory was the Busch's spokesman on WBRC. (Shaf Gregory collection.)

Even the nightly news was not safe from sponsor intrusion. Joe Langston says that after delivering a few news stories, he would go right into a Volkswagen commercial and then back to the news, without so much as a fade to black between them. (Joe Langston collection.)

The 1950s were the heyday of trading stamps, and WSGN pitched its lot with the yellow Top Value brand. Bill Bolen and Annette Griffis appear happy enough about it, but Toppie, the plaid Top Value Stamps elephant mascot, looks like he just got caught counterfeiting S&H Green Stamps. (Bill Bolen collection.)

Benny Carle had sponsors lining up to be promoted on his show. Here cowpoke Benny pours himself a heaping helping of one of the lucky ones, Foremost's brand of cow juice. (Benny Carle collection.)

Bozo the Clown is full of pep as he extols the virtues of Chocks, the "pillow-shaped" children's vitamin tablets. (Ward McIntyre collection.)

Instead of a single sponsor, Pat Gray promoted the whole of downtown Birmingham in this *c.* 1964 campaign. (Pat Gray collection.)

Like Rosemary, Bette Lee did not normally position herself as a "happy homemaker" type, but when sponsor Western Supermarkets called, she made herself at home on the kitchen range. (Bette Hansen collection.)

Remember the good old days when one could drop in to the downtown or Vestavia Sears store on a Saturday morning and rub elbows with Sergeant Jack or Bozo? (Author's collection.)

As mentioned earlier, Neal "Sergeant Jack" Miller got his name from his primary sponsor, Jack's Hamburgers. Miller passed away in July 2002. (Jean Spradlin-Miller collection.)

At some points, Cousin Cliff's *Popeye Show* had no fewer than 18 sponsors within an hour-long show. Consolidated Dairies was part of the herd. You are a true Birmingham kid if you remember the cow statues on the company's front lawn. (Cliff Holman collection.)

Five-and-ten chain V. J. Elmore was the official Birmingham sales agent for the vast line of *Romper Room* toys. Both Miss Jane and Miss Carol made numerous personal appearances at Elmore stores. (Jane Hooper collection.)

Now, who wouldn't buy a bottle . . . or six pack . . . or entire case of Pepsi when someone like Pat Gray was selling it? (Pat Gray collection.)

Six

A Lot Like Your Family

It was WBRC's promotional department that actually used the slogan "The Channel 6 Family: A Lot Like Your Family" to promote its lineup of local programming. The fact was that all of the local stations used the same personalities for so many different things that people did come to feel they knew them as well as they knew their Uncle Donnie or Aunt Laura. There was a less defined boundary between news and entertainment, and often the same people could be found handling both.

For example, Harry Mabry's reputation as a newscaster was unquestioned, but that did not prevent him from assuming master of ceremonies duties for the *Sunday Show*, a hodgepodge of acts that proved Bob Hope's contention that "when vaudeville died, television was the box it was buried in." Harry gave equal treatment to all the wannabes who made it onto the air, commenting on each and every act, "Mighty fine, mighty fine." In 1969, when Mabry left WBRC to start WHMA, Channel 40, in Anniston (and took Cousin Cliff Holman away from Channel 13 to help him do so), Joe Langston stepped effortlessly into the *Sunday Show* shenanigans.

One of the few television personalities who stuck to only one thing was Edward Burns, also known as Country Boy Eddy. Besides making local celebrities of Eddy and some of his regular cast members (such as the inimitable Zeke the Hayseed and piano player Mr. Wonderful), the *Country Boy Eddy Show* was also the training ground for later celebrities of national fame, including Tammy Wynette (who was, at the time, working as a hairdresser in Midfield) and young banjo pro Wendy Holcomb, whose pigtails kept the beat as she bounced around like a Jill-in-the-box.

In this final section, we will revisit many of these old friends of ours and close with a look at some of their reunions that have been held over the past few years. Most of them are completely retired from the broadcasting business now, but are still grateful to their loyal fans who made their careers possible. Of course, there is a high turnover in local television personalities today, and your favorite on-air talent might be gone tomorrow in search of a better position in another television market. For better or worse, the talents we are about to revisit for our farewell stubbornly stuck with their stations through it all, and in the process became local legends. To them, we can all say, "Thanks for being there when we needed you."

Along with news, talk shows, children's programs, and commercials, the Birmingham television stations even occasionally tried out the game show format. (Rosemary Lucas collection.)

Tom York began his long-running and highly successful *Morning Show* on Channel 6 in 1957. For one of his earliest programs, he soggily broadcast from the rainy Alabama State Fair. (Bill Bolen collection.)

Jim Lucas had a succession of cohosts for his *Teen Time Dance Party* show on Channel 13; here he is teamed with Bette Lee (left), who much preferred more serious journalism. The official rules for the show stated that no shorts could be worn on the program. Apparently no one told that to this unidentified girl. (Rosemary Lucas collection.)

Channel 13 entered the musical sweepstakes with *The Music Makers*, an eclectic collection of talents led by "Professor" Huel Murphy, who also played piano for the station's daily *Snicker Flickers* program of old silent movies. (Cliff Holman collection.)

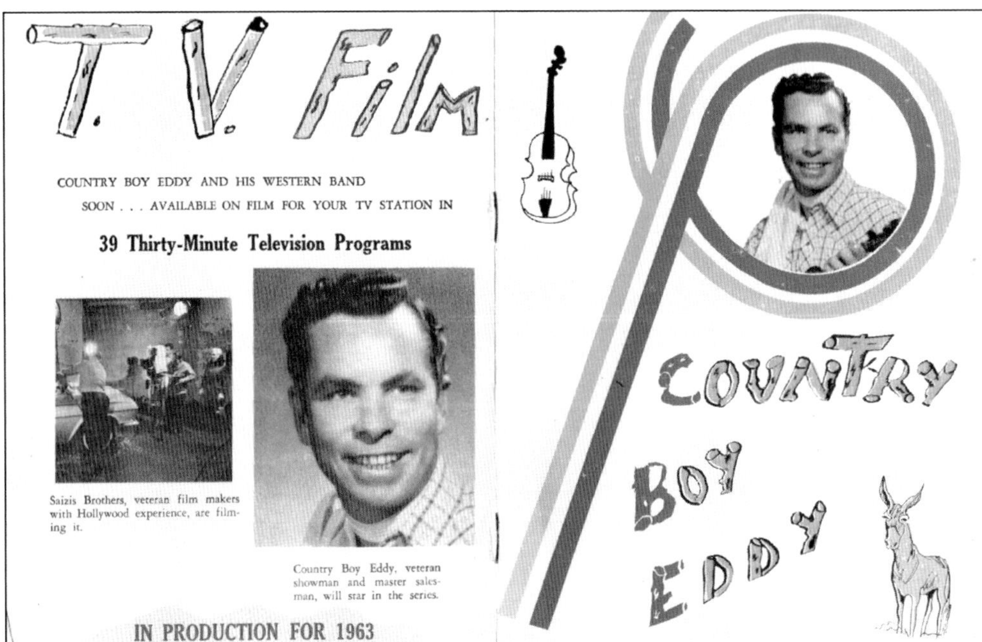

Country Boy Eddy (Burns, that is) and his band appeared on the *Morning Show* in 1957 after a period as a Saturday midnight fixture on Channel 13. Soon Eddy would have his own long-running early morning program on WBRC and would even film episodes for syndication across the South. (Edward Burns collection.)

Bette Lee and Jim Lucas compared flashing their pearlies with those of flamboyant pianist Liberace when he visited Channel 13 in the late 1950s. (Bette Hansen collection.)

Lee Thornberry, Miss Alabama of 1958, chats with Cliff Holman on a WAPI talk show. Children were sometimes surprised to see "Cousin Cliff" out of his nautical attire and "acting like a normal person," as he puts it. (Cliff Holman collection.)

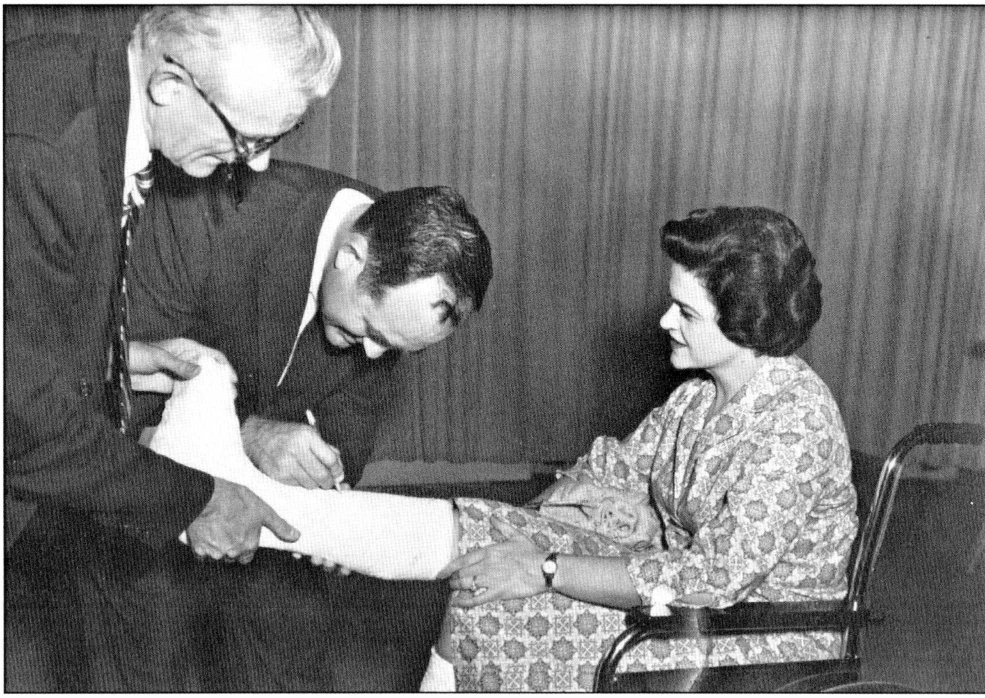

On election night 1958, John Patterson (center) was at WBRC, awaiting the returns. After the results were in, the first place he signed his name as "Governor John Patterson" was on Evelyn Allen's leg cast. Roy Marcado is helping the governor get a leg up on Evelyn. (Evelyn Allen collection.)

Channel 13 employee Norfleet Chenoweth tried out this photo opportunity board promoting Kellogg's newest star, Huckleberry Hound, c. 1959. (Norfleet Chenoweth collection.)

When famed baritone Vaughn Monroe came to Birmingham for the grand opening of Eastwood Mall in 1960, he dropped by WAPI to visit Cliff Holman's talk show. (Cliff Holman collection.)

Continuing its association with Eastwood Mall, in 1962, WAPI-TV and WAPI Radio broadcast a remote featuring, from left to right, two unidentified girls, Cousin Cliff, Dick Breit, Bette Lee, and an unidentified fan. (Dick Breit collection.)

Meanwhile, back at the *Morning Show* ranch, Tom York was holding a confab with Miss Kitty (Amanda Blake) and Doc (Milburn Stone) of *Gunsmoke*. Marshal Dillon must have been out killing a villain. (Tom York collection.)

Dennis the Menace (Jay North) strikes terror into the hearts of all adults during his visit to Channel 13 in 1963. The station managed to survive. (Norfleet Chenoweth collection.)

Neither Tom York nor Pat Gray can remember just why they won this trophy. It couldn't have had anything to do with Pat's legs, could it? Naaahhh. One amazing fact is that Pat's husband's name was Tom York, but he was a different one than her television costar. (Bill Bolen collection.)

After Bette Lee moved along to other projects, Rosemary Lucas filled the cohost seat next to Dick Breit on *Midday* for many more years. Above is the indefatigable duo's standard publicity shot. Below is a good view of the later *Midday* set, as Nanette Griffin of Sears describes the store's latest fashions for men and boys. (Rosemary Lucas collection.)

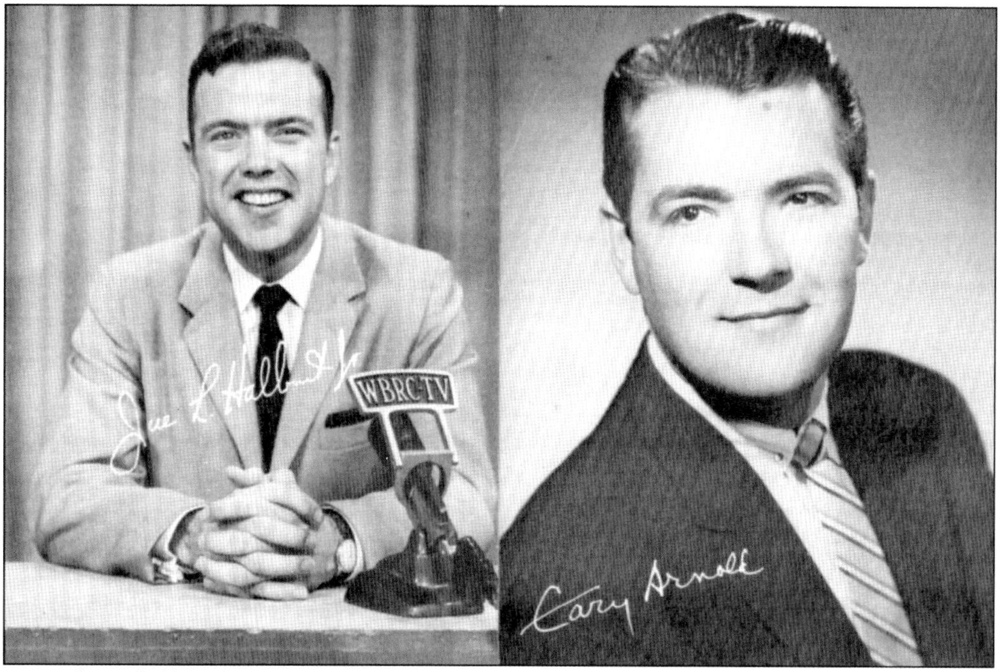

Joe Halburt (left) had a very short career at WBRC in the early 1960s. Gary Arnold (right) stuck around a bit longer, cohosting the women's game show *Petticoat Partyline* with Shaf Gregory. (Author's collection.)

Viewers never seemed to tire of beauty pageants, so for a few years, Ron Carney and Rosemary hosted the Miss Channel 13 contest. (Rosemary Lucas collection.)

Even though the *Morning Show* had been on the air since 1957, it really seemed to hit its stride around 1960. Notice that the Channel 6 logo still carried the CBS eye in this poster. (Pat Gray collection.)

Tom York and Pat Gray (plus many of Pat's successors as *Morning Show* sidekick) made a daily ritual out of exercising to Robert Preston's famed "Chicken Fat" workout song. (Pat Gray collection.)

Channel 13's contribution for early risers was *TV Partyline*, hosted by Dick Breit. He thinks the show might have been a bigger hit if it had used the title he suggested, *Breit and Early*. (Dick Breit collection.)

For one *Morning Show* broadcast, Tom York and sidekick Becky Finley (center) staged a genuine luau, courtesy of the Luau Restaurant near Eastwood Mall. Tom reports that the whole cast and staff got smashed on Polynesian mixed drinks before the show was over. (Author's collection.)

Although each started as a decorative "weather girl," both Pat Gray (left) and Rosemary (right) eventually found more rewarding work as serious journalists on Channels 6 and 13 respectively. (Author's collection.)

German-born Vera Alexander joined the *Morning Show* as Tom York's latest sidekick in 1966. (Author's collection.)

WBRC's Friday-night movie was originally hosted by Marynell Moore. When she left the station, the show was converted into *Home Theatre*, and these rare photographs document the filming of the title sequence at the Alabama Theatre, with new host Pat Gray selling tickets and oozing charm in the box office. (Pat Gray collection.)

Kaleidoscope was a quarterly WBRC program presented by Junior Achievement. This photograph identifies the tambourine-slapping singer as Debbie Garner and the guitar player at right as Chris Wright, but leaves the others unnamed. (Author's collection.)

While not clowning around as Bozo, Ward McIntyre could be found hosting horror movies as "Bela LaGhosty" (left) or introducing country music television shows as "Sad Sam" (right). (Ward McIntyre collection.)

As we have seen earlier, Fannie Flagg served in many different capacities at WBRC before her national fame as an actor and author. This publicity shot captures her more sedate side. (Fannie Flagg collection.)

Here come the *Morning Show*'s Three Musketeers in 1964: from left to right are Joe Langston, Fannie Flagg, and Tom York. (Joe Langston collection.)

During a period when Tom York had to be absent from the *Morning Show* for several weeks due to ulcer surgery, the show was carried on by this formidable WBRC stock company of talent. Pictured here are, from left to right, (standing) Joe Langston, Pat Gray, Ward McIntyre, and Harry Mabry; (seated) Benny Carle and Fannie Flagg. With a cast like this, how could the show not have gone on? (Joe Langston collection.)

Birmingham's Veteran's Day parade, said to be the oldest in the nation, was always a big television event. Above, Joe Langston described the parade for Channel 6 from atop the Tutwiler Hotel marquee in 1967. Below, Hugh Smith tackled commentary chores at the 1977 parade. (Above, author's collection. Below, Hugh Smith collection.)

Many people fondly (or not so fondly) remember Channel 42's live wrestling matches, announced by Sterling Brewer and sponsored by the hard-selling Taft Epstein of Epp's Jewelers. This artwork, incidentally, is by WBMG art director/*Sgt. Jack Show* puppeteer Howard Cruse. (Author's collection.)

This set, used by Country Boy Eddy for years, is today on display in the Alabama Music Hall of Fame at Tuscumbia. (Edward Burns collection.)

This advertisement was part of WBRC's "A Lot Like Your Family" campaign of 1976. Country Boy Eddy finally hung up his cowboy hat on the last day of 1993 but has continued making personal appearances ever since. (Author's collection.)

Former presidential first brother Billy Carter (center) dropped into the barn to share some cornpone jokes with Bill Bolen (left) and Country Boy Eddy in 1981. (Edward Burns collection.)

In 1980, Channel 13 took viewers back to the 1950s with a contest to win this vintage car. Seated in the automobile are, from left to right, Tommy Charles and James Spann; standing are John Ed Willoughby, Scott Palmer, Pam Huff, Wendell Harris, Rosemary, and Ken Snow. (Wendell Harris collection.)

The "Channel 6 Family" *c.* 1978 was made up of, from left to right, (first row) Donna Hamilton, Pat Gray, and Country Boy Eddy; (second row) Bill Bolen, Herb Winches, Tom York, and Joe Langston. (Bill Bolen collection.)

In 1982, the *Morning Show* reunited the ladies who had served as Tom York's sidekicks. Pictured here are, from left to right, Jeannine Johnson (1957–1958), Sarah Enfinger (1965), Pat Barnes (1962–1963), Sharon Brown (1963), Colleen Bates (1980–on), Tom, Evelyn Merriam (1972–1974), Pat Gray (1958–1962), Sharon Lacey (1967–1971), exercise instructor Mary Lowery, Phyllis Oliver (1975–1976), Becky Finley (1965), and Donna Hamilton (1977–1980). Absent were Fannie Flagg (1964), who was filming an episode of *The Love Boat*, and Vera Alexander (1966), who could not attend. (Tom York collection.)

From left to right, Country Boy Eddy, Benny Carle, Harry Mabry, and Cousin Cliff Holman swapped stories and lies during Channel 13's 50th anniversary party in May 1999. (Tom York collection.)

WBRC's 50th anniversary reunion in 1999 brought together, from left to right, Bill Bolen, Tom York, Harry Mabry, Scott Richards, Evelyn Allen, Ward McIntyre, Joe Langston, Oscar Michaels, and Country Boy Eddy. (Evelyn Allen collection.)

For our fade-out, we take a look at how Birmingham television might look in the Twilight Zone. After Pat Gray moved to Panama City, Florida, in 1979, she took a job with that city's Channel 13 and finally got to use her real last name. This just somehow looks bizarre to her legions of loyal Birmingham fans. (Author's collection.)